Luís Paulo F. Romão Silva

Nutritional enrichment of cupcakes from coconut mesocarp

Luís Paulo F. Romão Silva

Nutritional enrichment of cupcakes from coconut mesocarp

Authors Luís Paulo Silva Alfredina dos Santos Araújo Everton Vieira da Silva Gilcean Silva Alves Semirames Silva

ScienciaScripts

Imprint

Any brand names and product names mentioned in this book are subject to trademark, brand or patent protection and are trademarks or registered trademarks of their respective holders. The use of brand names, product names, common names, trade names, product descriptions etc. even without a particular marking in this work is in no way to be construed to mean that such names may be regarded as unrestricted in respect of trademark and brand protection legislation and could thus be used by anyone.

Cover image: www.ingimage.com

This book is a translation from the original published under ISBN 978-613-9-64105-5.

Publisher:
Sciencia Scripts
is a trademark of
Dodo Books Indian Ocean Ltd. and OmniScriptum S.R.L publishing group

120 High Road, East Finchley, London, N2 9ED, United Kingdom
Str. Armeneasca 28/1, office 1, Chisinau MD-2012, Republic of Moldova, Europe
Printed at: see last page
ISBN: 978-620-7-71994-5

SUMMARY

The coconut palm (Cocos *nucifera L.)* is one of the most widespread natural fruit trees on the planet, occurring on practically every continent, and has significant socio-economic value. As well as being a source of fibre, the mesocarp is also rich in minerals, proteins and has properties that can be used in the food industry as an important food supplement in the human diet, especially due to its fibre, protein, sugar and fat content. The aim of this study was to obtain and characterise coconut mesocarp bran and add it to the nutritional enrichment of cupcakes in different proportions. The bran was obtained after drying the mesocarp in natura at 70 °C for 48 hours and grinding it in a knife mill. The cupcakes were made in three different formulations (standard, 1% and 3%). The samples were subjected to physico-chemical analyses of moisture, ash, pH, acidity, proteins, lipids, fibre and total soluble sugars. They were also subjected to hygiene and health assessment (coliform at 45^0 C, coagulase positive staphylococcus, *Salmonella* sp, moulds and yeasts, *Escherichia coli*). The sensory analysis was carried out with 146 untrained tasters, samples were presented in individual booths, in a monadic and randomised manner, checking the attributes of appearance, colour, flavour, aroma, texture, overall acceptance by the affective method, using the product acceptance test. The cupcake samples presented satisfactory sanitary conditions in relation to the hygiene and health assessment, while in relation to their chemical composition they presented protein (4.90%), lipid (5.93%) and fibre (1.13%) levels, which could make them a food rich in fibre and protein. The sensory analysis showed that the formulation with 3% mesocarp bran was the most acceptable in terms of its attributes.

Keywords: Tropical fruit, waste, nutritional enrichment.

Summary

CHAPTER 1

Introduction

The coconut palm (Cocos *nucifera* L.) is one of the most widespread natural fruit trees on the planet, occurring on practically every continent and with significant socio-economic value. Due to its dispersal and adaptability, its cultivation and use has spread significantly throughout the world, with a wide variety of products, both *fresh* and industrialised. Over time, coconut production has grown considerably in the North and Northeast regions. This sharp increase came at the end of the 1990s with an increase in the area under cultivation and the types of coconut palm species planted, which are dwarf and hybrid coconut palms that are used to produce coconut water (EMBRAPA, 2011).

The fruit tree originated in the tropical and subtropical islands of the Pacific Ocean, with Southeast Asia as its main centre of origin and diversity, and its spread over time to Latin America, the Caribbean and tropical Africa (FOALE, 2010).

Every part of the plant has a purpose, such as the root, stem, leaf, inflorescence and fruit, which are used for craft, food, nutritional, agro-industrial, medicinal and biotechnological purposes, among others. The main purpose and utility in the country is the industrialisation and use of coconut water. Brazil is currently the world's fourth largest coconut producer with approximately 2 billion fruits harvested, accounting for 80 per cent of South America's total production (EMBRAPA, 2014; FAO, 2014).

Coconut shells (Cocos *nucifera* L) are an agricultural waste product with great potential for reuse, but with little action in this regard. It is currently used more in construction, in the production of Portland cement raw materials and ornamental garden frames, and 85% of this waste is considered rubbish for no purpose. Its fibre has a long degradation period, around nine years to decompose in the environment (CARRIJO, 2002).

Coconut waste generates a significantly high and growing volume of unusable material. One estimate suggests that this amount of coconut waste is approximately 6.7 tonnes of husk per year and is deposited in dumps and landfills, causing serious environmental problems with the disposal of this material (ROSA, 1998).

Taking into account the processing of the coconut to remove the water and pulp, a large amount of waste is generated, and the mesocarp can be transformed into bran for addition to foods with a low fibre content. As a result, the demand for foods rich in fibre or sources of fibre has gradually increased, as well as being healthy and helping to maintain the body.

Therefore, in Brazil, a country characterised by its agro-industrial economy and with the potential to become one of the largest producers of cellulose fibre, the development and reuse of fibre for food purposes could lead to economic growth for producing regions and a reduction in the waste that is dumped in rubbish dumps and landfills (PEREIRA, 2012).

Bakery products are one of the most consumed in the country and the world, and cake is a popular food. The term cake is used to refer to products that are characterised by formulations based on wheat flour, sugar, whole eggs and fat, and other components where each ingredient plays an important role in the structure and quality of the food (MATSAKIDOU; BLEKAS; PARASKEVOPOULOU, 2010; RAMOS, 2012). This product gave rise to the cupcake, which is a small cake served as a snack at school or in the afternoon.

CHAPTER 2

Objectives

2.1 General Objective

To evaluate the reuse of coconut mesocarp in the nutritional enrichment of cupcakes.

2.1.1 Specific objectives

- Obtain coconut mesocarp flour;
- Characterise the bran through physico-chemical and microbiological analyses;
- Making enriched cupcakes from coconut mesocarp; evaluating the enriched cupcakes based on physicochemical and microbiological analyses.
- Sensory analysis of the food using affective methods in accordance with CAAE approval n° 59890316.2.0000.5575 from the Ethics Committee of the Federal University of Campina Grande.

CHAPTER 3

Bibliographical review

3.1 Characterisation of Coconut

The coconut palm has a simple trunk, erect or slightly curved, with leaves up to three metres long and yellowish-green leaflets. Its clusters, which are approximately one metre long, are made up of white flowers and large ovoid-shaped fruits, 25 cm long and 15 cm in diameter. This palm tree is of great economic importance throughout the world, from which it is possible to extract pulp (albumen), water for food, leaves for agricultural purposes, handicrafts, and also fibres for the construction industry (MUSEU NACIONAL, 2015).

The coconut (Coco *nucifera L.)* is botanically classified as a fibrous drupe, made up of: the epicarp or smooth epidermis, the layer that surrounds the mesocarp; the mesocarp, the thick, fibrous layer (shell); and the endocarp, the woody layer that surrounds the seed, making it very hard after ripening. Between the endocarp and the solid albumen is a thin layer, light in colour in immature fruit and brown in ripe fruit, called the integument. The albumen or solid endosperm in ripe fruit is a fleshy, white, oily, fairly thick layer, and in immature fruit, depending on its stage of ripeness, it has a semi-solid (gelatinous) consistency (EMBRAPA, 2003).

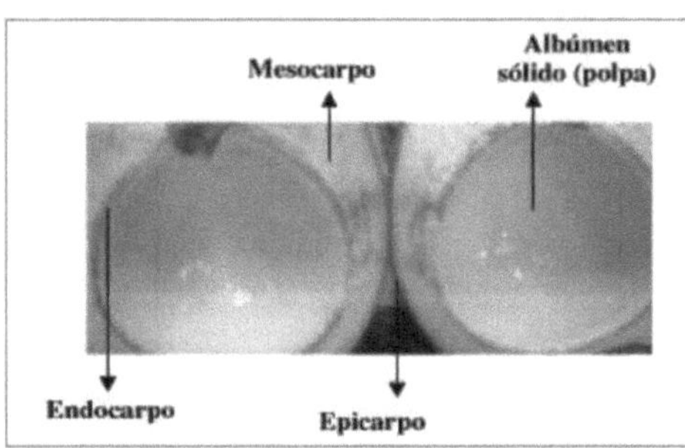

Figure 1: Coconut fruit.
Source: Santana (2012).

The fruit of the coconut palm is rich in B vitamins (B-i, B_2 and B_6) and vitamin C. It is also a source of fibre and minerals such as calcium, iron, magnesium, phosphorus, potassium and sodium. Coconut water is made up of glucose and minerals such as sodium, potassium and chlorine and is used for hydration during the summer period. The pulp is made up of lipids, proteins and fibres and is used to extract coconut milk, which is used in food products by the food industry in various segments (FRUTAS DO BRASIL, 2015). Coconut production has been growing considerably, generating foreign currency, employment, income and food. The country's production was 2.8 million tonnes of fruit, making it the top producer in South America. Although coconut cultivation is being boosted in various regions of the country, the largest crops are concentrated in the north-east and north, accounting for 70% of national production. This production is favoured by tropical climatic conditions, as can be seen in tables 1 and 2 (EMBRAPA, 2013).

Table 1: Production, harvested area and productivity of the main coconut-producing countries in South America, 2012.

Country	Harvested Area (ha)	Production (1000 tonnes)	Productivity (1000 tonnes/ha)
Brazil	257.742	2.888.532	11,21
Venezuela	19.000	165.000	8,68
Guyana	17.000	80.000	4,71
Colombia	12.900	102.000	7,91
Ecuador	3.300	20.000	12,01
Peru	2.472	29.687	6,06
Suriname	550	4.000	7,27
French Guiana	65	350	5,38
South America	313.029	3.289.569	9,6

Source: FAO, 2014.

Table 2: Area planted with coconut trees, coconut production and productivity in the regions of Brazil in 2012.

Regions of Brazil	Planted Area (ha)	Production value (thousand fruits) production (R$)
North-East	208.977	1.345.962 579.899
North	27.314	252.406 104.676
South East	20.471	315.714 182.714
Centre-West	2.752	37.190 27.666
South	223	3.082 2.722

Source: IBGE, 2014

The increase in coconut production has become a natural trend, resulting in a high generation of solid waste (husk). The waste consists of the mesocarp, the fibrous part of the fruit. Unlike the shell of mature coconuts, which is used in the construction industry, as fuel for boilers and in agriculture, there is no suitable technology for reusing green coconuts, except in the production of agricultural substrate (MATTOS, 2011).

2.2 Healthy eating

In search of healthier foods, with benefits for good health, the concept of functional foods emerged in the 1980s in Japan, which in addition to providing basic nutrition to the individual, promote health through the potential of the food ingested in the diet. These food ingredients that provide health benefits include disease prevention and treatment. The benefits of eating these foods ensure the maintenance of well-being, modulating the body's physiology and promoting hypocholesterolaemic, hypotensive, atherosclerosis risk reduction and anticancer effects (BASHO, 2010; OLIVERA,

2002).

This trend towards functional foods is growing at a very high rate, with the population wanting healthier foods with beneficial substances, containing probiotics, prebiotics, antioxidants, immunopeptides, isoflavones and other substances. They have the ability to combat problems such as insomnia, stress, constipation and other illnesses, replacing medicines in the fight against them. Some products have been well received by the consumer market, such as yoghurts and breads that are used to improve the digestive tract (BNDS, 2015).

In order to lead a healthy life, people have changed their habits and the consumption of functional products, which help to combat certain chronic or non-communicable diseases, has increased due to the nutrients present in food and the benefits they bring to the body's physiological functions (OLIVEIRA, 2008).

3.   3Reutilisation

In recent years, special attention has been paid to minimising or reusing the solid waste generated by different industrial processes. Waste from the food industry involves significant quantities of peels, stones and other fruit components. As well as being a source of organic matter, these elements serve as proteins, enzymes and essential oils that can be reused and recovered (COELHO, 2001).

The growing consumption of green coconuts and their natural introduction for the industrialisation of coconut water has gradually increased, consecutively generating a large production of waste (coconut shell), which corresponds to around 85% of the weight of the fruit (COELHO, 2001). Coconut mesocarp is characterised by its hardness and the durability of the lignin content in its fibres, when compared to natural fibres. It degrades slowly, taking more than 8 years to completely decompose. As minimising the generation of this waste would mean a reduction in the associated production activity, its use has become a necessity (CORRADINI, 2009).

The mesocarp fibre has the function of reinforcing materials, thanks to its rigidity and high resistance, giving products durability (ROSAS, 2009). As well as being a source of fibre, the mesocarp is also rich in minerals, proteins and has properties that can be used in the food industry as an important food supplement in the human or animal diet. It is extracted by crushing the mesocarp, drying it and then crushing and grinding it. After this process, it is introduced into food as a dietary supplement due to its richness in nutrients (FERREIRA, 2010).

Physico-chemical characterisation studies of the mesocarp of green coconuts are scarce, which highlights the importance of studies of this nature in order to make better use of the plant and its industrial waste, and to provide nutritional enrichment to the foods that can be produced with its addition.

3.4 Drying Principle

Drying is one of the unit operations used to remove water from a product by evaporation or sublimation, using heat under controlled conditions. The purpose of this operation is to preserve food by reducing its water activity. In recent decades, science and technology have endeavoured to improve new systems in the area of food preservation, and these efforts have made it feasible to dehydrate large varieties of products for commercial purposes. The fundamentals of drying are based on the theory of mass and heat transfer together with mass and energy balance. This process removes a large part of the water in the food (KAJIYAMA; PARK, 2008; ATHIÉ *et al*, 1998).

According to Park et al. (2006), during drying it is necessary not only to supply heat to evaporate the moisture from the material, but also a moisture sorber in order to remove the water vapour formed from the surface of the material to be dried; it is this process of supplying heat from the hot source to the moist material that will promote the evaporation of the water from the material and then the mass transfer will drag the vapour formed. It can be seen that two phenomena occur simultaneously during the process when a moist solid is subjected to drying:

- Transfer of energy (heat) from the environment to evaporate surface moisture; this transfer depends on external conditions of temperature, air humidity, air flow and direction, exposure area of the solid (physical form) and pressure;

- Transfer of mass (moisture) from the interior to the surface of the material and its subsequent evaporation due to the first process; the internal movement of moisture in the solid material is a function of the physical nature of the solid, its temperature and moisture content.

The drying process must take place in a controlled manner, so that it can occur uniformly, avoiding high gradients of humidity and temperature inside the material, which can cause the product to lose its quality. Knowing that the effects of drying alter the physical and chemical properties and sensory characteristics of the product and, in turn, affect the process of heat and mass transfer, it is essential to know its effects and how to control them (FARIAS et aL, 2002).

Among the by-products generated by coconut processing is the husk (mesocarp), which when ground results in a brown-coloured bran with a characteristic smell. According to Brazilian legislation (BRASIL, 2005), bran is the product resulting from the processing of cereal grains and/or legumes, consisting mainly of husk and/or germ, and may contain parts of the endosperm.

Evaluating the composition and availability of nutrients and energy in alternative feeds is important for formulating nutritionally and economically viable rations. Research into these feeds aims to make the best use of nutrients during ingestion, avoiding deficiencies or excesses, which helps both to reduce costs and to reduce the excretion of nutrients into the environment (ROSTAGNO, 2007).

The chemical composition of food, determined through bromatological analyses, only describes the nutrients present in the food, and it is also necessary to know the digestibility of its nutrients in order to better express their absorbed portions (PARSONS, 1996).

Studies show that cereal brans (rice, wheat, rye, oats, barley) are being added to make new products or for nutritional enrichment, as these residues contain a significant amount of protein, fibre, lipids and minerals. The use of rice bran is becoming widespread in the production of biscuits and cakes for nutritional enhancement (LACERDA,

2008).

Rodrigues (2010) used cassava bran, which is a by-product of starch processing, to make starch biscuits at different levels of substitution of starch for bran, obtaining high protein and fibre values.

3. 5Fibres

Fibre is the set of substances that make up the skeleton of plants, without which no plant could remain upright, as it forms the walls of each of its cells. It is now recognised that dietary fibre (DF) refers to a complex material of plant origin that is resistant to digestion by enzymes in the human intestinal tract. The term dietary fibre is generic and covers a range of chemically defined substances with peculiar physicochemical properties and individual physiological effects (INSUMOS, 2016).

Dietary fibre, formerly known as crude fibre, is material that cannot be digested by the human or animal body. They are insoluble in acid and base diluted under specific conditions. Crude fibre has no nutritional value, but it provides the necessary tool for the peristaltic movements of the intestine and absorbs free radicals in the intestine (CECCHI, 2003; DAMODARAN, 2010). The fact that fibre is not hydrolysed by digestive enzymes does not mean that it is not partially degraded and metabolised. Between 10 and 80% of it undergoes a fermentation process in the colon, giving rise to compounds that the body absorbs and metabolises (ÓRDONEZ, 2005). Dietary fibre is nutritionally important because it maintains the normal functioning of the gastrointestinal tract. Its presence in food induces satiety at mealtimes (DAMODARAN, 2010).

In recent years, dietary fibres have been attracting interest from the scientific community due to the number of effects this nutritional component has on a wide variety of pathologies, including cancer, hypercholesterolaemia, diabetes and hypertension. Fibre has positive modulatory effects on some cells in the immune system and on the metabolic and biochemical processes associated with exercise, influencing the performance of athletes, improving their performance and recovery from daily training sessions (INSUMOS, 2016).

Table 3: Classification of dietary fibres.

Starch polysaccharides	Soluble non-starch polysaccharides	Polysaccharides insoluble starches	Polyphenols and other compounds associated with the cell wall
Resistant starch (types I and II) Retrograde starch	Gums Mucilages Pectins	Cellulose Hemicelluloses	Lignin Cutin Tannins Suberin Phytates Protein minerals Ca^{2+}, K^+, Mg^{2+}
Soluble dietary fibre		Insoluble dietary fibre	

Source: INSUMOS, 2016.

3.6   Ready-made cake

Among bakery products, the ready-made cake has become increasingly important in terms of consumption and commercialisation in Brazil, due to its easy production logistics and its technical development, which has made it possible for the bakery industry to change from small to large-scale production (MOSCATTO, 2004).

Consumed mainly for breakfast and as a snack for children at break time and in the afternoon, there is greater demand and market growth, causing companies to expand their flavour and size options. And to help boost the segment's potential, supermarkets are reserving more space on their shelves for ready-made cakes every day (MORAIS et al, 2014).

According to Abima (Brazilian Association of Pasta, Bread and Industrialised Cakes Industries), consumers regardless of social class, together with the growing demand for more practical products, explain the good performance of the industrial cakes sector. The reason is that many families no longer have time to prepare homemade cakes and end up opting for the convenience of buying ready-made products, which are now available in much greater variety on the shelves of retail outlets (ABIMA, 2013).

According to Abimapi (2016), the sale of industrialised cakes in Brazil amounted to 37,30 thousand tonnes in 2015, bringing in around 0.85 billion reais, making Brazil 6th° in the ranking of countries in terms of sales. Studies show that there is a great demand for functional foods, and the addition of these compounds to cakes increases the demand for this type of food, characterising a concern for improving quality of life (ZANINI, 2016).

CHAPTER 4

Material and method

The preparation and development of the bran and cupcakes, as well as the physical, chemical, microbiological and sensory analyses were carried out in the laboratories belonging to the Academic Unit of Food Technology - UATA and the Technological Vocational Centre - CVT, of the Centre for Agri-Food Sciences and Technology - UFCG.

4.1   Material

4.1.1 Raw materials
4.1.2 Processing coconut bran

The coconut was purchased from the trailers at the Municipal Bus Station of Pombal, located on the banks of the BR 230 highway, and transported to the laboratory at the Technological Vocational Centre - CVT, of the Centre for Agri-Food Sciences and Technology - UFCG. The samples were washed, sanitised in a bucket containing 115 ppm sodium hypochlorite solution for 10 minutes, then the mesocarp was separated from the endocarp and dried until the moisture was completely removed in a New Lab N1040 air circulation oven at a temperature of 70°C ± 2°C for 48 hours. After drying, the mesocarp samples were ground in a knife mill with 4 stainless steel blades and a 200 mesh sieve. At the end of processing, a brown bran with a characteristic smell was obtained, which was packed in sterilised polypropylene plastic jars with a capacity of 500g (Figure 4), sealed and kept at room temperature until further use.

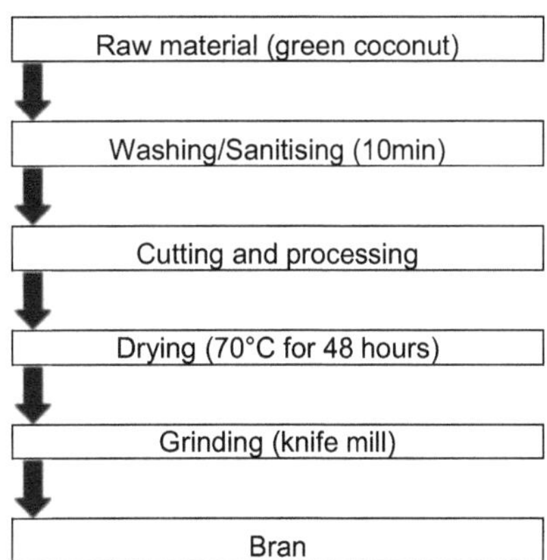

Figura 2: Flowchart illustrating coconut processing.

Source: Prepared by the author.

Figura 3: Samples submerged in hypochlorite and sanitised.

Source: Own authorship.

Figure 4: Coconut mesocarp bran.
Source: Own authorship.

4.1.3 Making and processing cupcakes

The raw materials used to make the cupcakes with coconut mesocarp were: unleavened Primor wheat flour, Puro Sabor margarine, Parari crystal sugar, Dr Oetker vanilla essence, Italac whole milk, Dr Oetker dry chemical yeast and coconut mesocarp bran processed in the laboratory.

4.1.4 Cupcake processing

For the preparation of the cupcake batter, special wheat flour, butter, sugar, whole milk, vanilla essence and dry chemical yeast from local shops were used. After defining the base formulation, the wheat flour was partially replaced with coconut mesocarp bran, and the cupcake batter was processed by hand in the following concentrations: Sample 1 - 99% wheat flour and 1% coconut mesocarp bran; Sample 2 - 97% wheat flour and 3% coconut mesocarp bran; Sample 3 - 100% wheat flour (control formulation).

Table 4: Ingredients used to make the cupcakes and their respective formulations.

Ingredients	Formulations		
	Control	1% mesocarp	3% mesocarp
(p/240g)			
Sugar	176g	176g	176g
Wheat flour	240g	237,6g	232,8g
Chemical yeast	5g	5g	5g
Mesocarp bran	og	2,4g	7,2g
Whole milk	250mL	250mL	250mL
Margarine	24g	24g	24g
Egg	2 units	2 units	2 units
Vanilla essence	10mL	10mL	10mL

Source: Prepared by the author.

First, the dry ingredients were mixed with the margarine and egg, then the milk and vanilla essence were added until a homogeneous dough was obtained. The dough was transferred to the paper and aluminium moulds and baked in the preheated oven for 25 minutes. After the cake had cooked, it was unmoulded and placed in polythene trays and covered with PVC film at room temperature for analysis.

Figure 5: Finished cupcake.
Source: Prepared by the author.

4.1.5 Physico-chemical parameters of coconut mesocarp bran (Cocos *nucifera L)*

The physicochemical characterisation was carried out on the coconut mesocarp bran, following the methodology described by IAL (2008). The following analyses were carried out: moisture content (%), ash (%), total lipid content (%), total acidity (mg/g), pH, protein (%), total fibre (%) as described by AOAC (1997), total sugars (IAL, 2008).

4.1.5.1 pH

Hydrogen potential (pH) was determined using the potentiometric method with a Lucadema model mPA bench pH meter, previously calibrated with pH 4.00 and 7.00 buffer solutions. Following method 017/IV of the Adolf Lutz Institute (2008).

4.1.5.2 Titratable acidity (TA)

50mL of distilled water was added to approximately 5g of bran, weighed previously in a 100mL beaker. After homogenising with the aid of a glass rod, it was filtered through qualitative filter paper into 125 mL erlenmeyer flasks. The acidity was determined by titrating these solutions with 0.1mol/L sodium hydroxide solution (NaOH) until a pH range of 8.2 - 8.4 was reached.

This technique is recommended for dark or strongly coloured solutions, where the equivalence point is determined by measuring the pH of the solution, following methodologies 311/IV and 312/IV described by the Adolfo Lutz Institute (2008). The results were expressed in grams of citric acid/100g of sample.

4.1.5.3Humidity   (%)

The procedure was carried out by direct heating in an oven at 105°C. The crucibles used for the analysis were dried in an oven at 105°C for 1 hour, then placed in a glass desiccator for 15 minutes. They were then weighed on an analytical balance. The weights were duly recorded. Next, 5g of each sample was weighed. The sample crucibles were placed in an oven at 105°C for 5 hours, then placed in a glass desiccator for 15 minutes and weighed again, repeating until a constant weight was reached (IAL, 2008).

4.1.5.   4Fixed mineral residue   (Ash)

The ash content (%) was determined by calcifying approximately 5g of the sample in a Quimis muffle furnace at 550°C (Figure 6) for 18 hours.

Figure 6: Murfla.

4.1.5.5Proteins   (%)

The protein content (%) was determined using the Kjeldahl method. The samples were prepared with 0.2g of coconut mesocarp bran, 1.5g of catalysts (potassium sulphate and copper sulphate) and 3mL of PA sulphuric acid, digested (New Lab NL 23-01 digester block) (Figure 7a) under gradual heating with a heating rate of 50°C until it reached 400°C. After the digestion process, the system was added 40 mL of distilled water, 5 mL of 63% sodium hydroxide and phenolphthalein as an indicator, and distilled in a Solab SL 74 nitrogen distiller (Figure 7 b). The distilled material was collected in a container with boric acid and indicators (methyl orange and bromocresol green) and then titrated with a 0.1 M hydrochloric acid solution. The result was expressed as a percentage (IAL, 2008).

Figure 7: Image of the Digester Block and Nitrogen Distiller.
Source: Own authorship.

4.1.5.6Lipids   (%)

The lipid content was checked using the direct extraction method in Sohlext (Figure 8) described by IAL (2008). To do this, around 5g of the sample was measured and hexane was added as a solvent in a Solab Sohlext apparatus. The system was heated for around 6 hours and then dried in an oven at 105 °C for 1 hour to remove the excess solvent and until the constant weight of the fat produced was reached.

Figure 8: Lipid extractor.
Source: Own authorship.

4.1.5.7 Total Soluble Sugars   (TSS)

Total soluble sugars were determined using the Antrona method described by Yemn and Willis (1954). Samples of bran were dissolved in water and filtered through filter paper. Aliquots were taken from the resulting filtrate and reacted with Anthrone in a bain marie at 100 °C. The end of the reaction was detected by acquiring a characteristic colour. Readings were then taken on an AAKER SP 2000 UV spectrophotometer (Figure 9) at a wavelength of 620 nm and the result was obtained using the straight line equation of the glucose standard curve and expressed as a percentage.

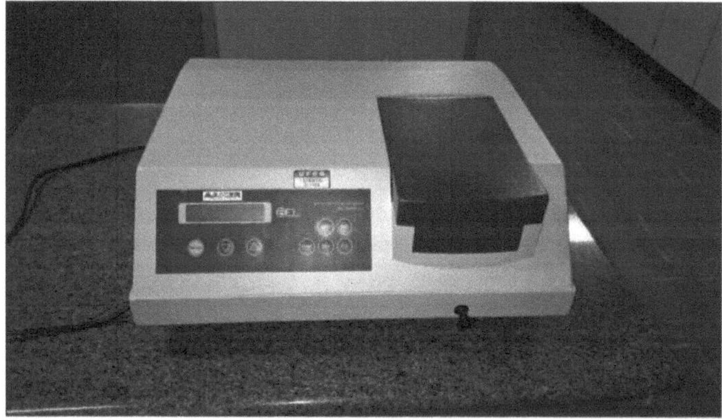

Figure 9: Spectrophotometer.
Source: Own authorship.

4.1.5.8 Fibres   (%)

The method used to determine the crude fibre content was the acid-base digestion method described by AOAC (1997), with modifications made by Pontes júnior (2012). In this methodology, two procedures were carried out to wash the sachets in acid and base, the first consisting of washing the sachet without a sample and the second with a sample. In the first procedure, after making TNT sachets (5 cm x 5 cm) and sealing them with a sealing press (Figure 10 b), they were identified and boiled in a 2.25 L acid solution in the SL118 Solab fibre determiner (Figure 10a) for fifteen minutes and then washed three times with boiling distilled water, the first time for five minutes and the other two for three minutes each. They were then boiled in a 2.25 litre basic solution for fifteen minutes, followed by the same washing procedure with distilled water as described above. This prior washing is done to environmentalise the bags, preventing any later errors in quantification. After washing, the bags were placed in an oven (105 °C) for sixteen hours, after which they were placed in a desiccator with tweezers for forty minutes and then weighed and the values of the empty bags recorded. In the second procedure, for each sample, 1g of the coconut mesocarp meal was weighed and added to the bags. The bags were sealed,

The sample was spread evenly inside the sachets and then transferred to the fibre determinator holder. The equipment containing 2.25 litres of acid solution at room temperature was closed. When the temperature reached 95 °C, a 30-minute timer was set. Once this time had elapsed, the acid solution was drained into an appropriate container for subsequent neutralisation and disposal, and the sachets were then washed three times with boiling distilled water while the apparatus was stirred to remove the excess acid solution, the first time for five minutes and the other two for three minutes. At the end of each wash, all the water in the equipment container was drained and replaced for the next wash.

2.25 litres of basic solution were added to the apparatus and the solution was allowed to heat up (95 °C), followed by a 30-minute timer. When the extraction was complete, the apparatus was switched off and the basic solution drained. The bags were then washed again with boiling distilled water as described above. The sachets were then removed from the apparatus support and distributed on stainless steel trays lined with paper towels and placed in an oven (105 °C) for sixteen hours. After this period, the sachets were removed and placed in a desiccator for an hour to balance the temperature and humidity, then weighed and the values recorded. The crude fibre content was obtained from the difference between the weight of the dry sachet with the sample after acid-base digestion and the weight of the dry sachet without the sample before digestion. The value obtained was multiplied by 100 to obtain the crude fibre content as a percentage.

The crude fibre value was obtained by calculating the equation.

$$\%FB = (PD - \frac{TARA}{PA})x100$$

Crude fibre = percentage of crude fibre in the feed;

PD = weight of sachet + sample (g);

Tare = weight of empty sachet (g);

PA = sample weight (g).

Figure 10: Samples for fibre determination and the apparatus used in the analysis.
Source: Own authorship.

4.1.6   Physico-chemical parameters of cupcakes.

The physico-chemical characterisation was carried out in triplicate. The methodology used by Pontes Júnior (2012) and IAL (2008) was followed. The following determinations were made: moisture content (%), ash (%), total lipid content (%), total acidity (mg/g), pH, protein (%), total fibre (%) as described by AOAC (1997).

4.1.6.1 pH

The methodology described in section 4.1.5.1 is followed.

4.1.6.2 Fixed mineral residue (Ash)

The methodology described in section 4.1.5.4 is followed.

4.1.6.3 Humidity (%)

The methodology described in section 4.1.5.3 is followed.

4.1.6.4 Total Titratable Acidity (ATT)

The methodology described in section 4.1.5.2 is followed.

4.1.6.5 Protein (%)
The methodology described in section 4.1.5.5 is followed.

4.1.6.6 Lipids (%)
The methodology described in section 4.1.5.6 is followed.

4.1.2.1 Fibres (%)
The methodology described in section 4.1.5.8 is followed.

4.1.7 Microbiological Parameters of Coconut Mesocarp Bran
For the microbiological analyses, the mesocarp bran sample (25g/sample) was diluted in 225mL of 0.1% peptone water and homogenised on a Nova Ética® orbital shaker (109/2TCM) at 180 rpm for 20 minutes (10' dilution[1]). Subsequent decimal dilutions were prepared using the same diluent. *Salmonella* sp, coagulase positive staphylococcus, coliforms at 35° and 45°C, moulds and yeasts were analysed from the first dilution.

4.1.7.1 Salmonella sp.
For the identification of Salmonella sp/25 g, the ÀGAR RAMBACH culture medium (Himedia®, Munbai, India) was used and incubated in a bacteriological oven at a temperature of 36±1°C for 48 hours with adaptations (SILVA, 2010).

4.1.7.2 Coagulase positive staphylococcus
To analyse coagulase-positive Staphylococcus, 0.1 mL of each selected dilution was inoculated onto the dry surface of Mannitol (Himedia®, Michigan, USA). Incubation was carried out in a bacteriological oven at a temperature of 36±1°C for 48 hours (SILVA, 2010).

4.1.7.3 Moulds and yeasts
For the analysis of moulds and yeasts, 0.1 mL of each selected dilution was inoculated onto the surface of Potato Dextrose Agar (Himedia®, Michigan, USA). The plates were incubated at 25°C for 5 days, according to the recommended methodology (BRASIL, 2003).

4.1.7.4 Coliforms at 35 and 45 °C
To identify the Coliform Group, each dilution was sown in three tubes containing Lauryl Sulphate Tryptose Broth (LST, Himedia®, Curitiba, Brazil) for the quantification of the presumptive test (MPN). Incubation took place in a bacteriological oven at 35±2 °C for 24 hours and those with turbidity or gas production collected in the inverted durhan tube were considered positive. To determine the confirmatory test for coliforms at 35 °C, the multiple

tube technique was used with three series of three tubes containing 2% Bright Bile Green Broth (Himedia®, Mumbai, India), incubated at 35±2 °C for 24 hours. From the positive tubes, tubes containing EC Broth (thermotolerant) were replica-plated and incubated at 45±1°C for 48 hours in a water bath with water circulation model Q-215M2 Quimis (BRASIL, 2003).

4.1.8 Sensory evaluation of the cupcakes.

In accordance with CAAE approval No. 59890316.2.0000.5575 from the Ethics Committee of the Federal University of Campina Grande, Cajazeiras Campus - PB, the test was conducted with 146 untrained tasters of both genders, using a 9-point verbal hedonic scale (STONE and SIDEL, 2004), with scores ranging from 1 (I disliked it very much) to 9 (I liked it very much). The sensory attributes adopted were appearance, colour, aroma, flavour, texture and overall acceptability (Appendix A).

The evaluators' intention to buy was also assessed in relation to the samples presented, using a five-point structured scale, where the judges gave scores from 1 to 5, ranging from "certainly would buy" to "certainly would not buy" as described in the sensory evaluation form in Appendix A.

The sample of each formulation was presented simultaneously to the evaluators in individual booths and served on disposable plates, randomly coded with a three-digit number, accompanied by the evaluation form, a cracker and a glass of water to cleanse the taste buds and remove residual flavour.

4.1.9 Statistical analysis

The data obtained in this study was subjected to Analysis of Variance (ANOVA), using a block design with repetition, where the mean and standard deviation were obtained. The results were then compared using Tukey's test at a significance level of 5%, using the ASSISTAT statistical programme version 7.7 beta.

CHAPTER 5

Results and Discussion

5.1 Microbiological characterisation of bran

The green coconut mesocarp bran sample was evaluated using microbiological tests after processing. The presence of coliforms in the temperature range of 35 to 45 °C, *Salmenella* sp, *coagulase* positive staphylococci, moulds and yeasts and *Escherichia coli* were investigated, as recommended by Brazilian legislation through RDC n. 12, of 2 January 2001.

The data relating to the microbiological analyses are shown in Table 5 below.

Table 5: Microbiological analysis of coconut mesocarp bran.

Parameters	Sample	VMP
Coliform at 35 °C (MPN/g)	<3	-
Coliform at 45 °C (MPN/g)	<3	5×10^2
Coagulase positive staphylococci (CFU/g)	$0{,}96 \times 10^2$	-
Moulds and yeasts (CFU/g)	Absent	-
Salmonella sp/25g	Absent	Absence
Escherichia coli	Absent	-

[1]VMP: Maximum permitted value, according to Brazil (2001).

[2] MPN: Most probable number.

[3] CFU: Colony forming unit.

As shown in Table 5, coliform levels at 45 °C were below those recommended by RDC no. 12, so this did not affect the quality of the bran to be added to cupcakes. As for the presence of *Salmonella* sp, coagulase positive staphylococci and moulds and yeasts, there were no traces of these microorganisms, which shows excellent sanitary conditions during the processing and packaging of the bran for the preparation of the product.

4.2 Physical and Chemical Evaluation of Coconut Mesocarp Bran

Table 6 shows the physico-chemical data obtained from the coconut mesocarp bran in terms of pH, acidity, moisture and ash parameters.

Table 6: Physico-chemical parameters for pH, acidity, moisture and ash of coconut mesocarp.

Sample	Parameters			
	Acidity (%)	Ash (%)	PH	Humidity (%)
Mesocarp bran	1,09 ±0,02	4,80 ±0,02	5,51 ±0.07	1,58 ±0,34

Averages of three repetitions followed by the respective standard deviations.

As described in Table 6 above, the average values obtained for the pH of the bran was 5.51, which is close to the data found by Rodrigues (2010) in a study of cassava bran (pH = 5.01) and Silva (2013) who found a pH of 5.90 for mesocarp bran from Mari.

Other studies have also found very similar data, as in the case of pregelatinised rice flour, where Ruiz et al (2003) found a pH ranging from 5.8 to 6.4. Therefore, the data from this study is in agreement with the fact that the bran obtained may not be considered to be up to the standards of commercial products for use in food.

As for acidity, the sample of coconut mesocarp bran showed different values to common flour products, such as cassava flour (2.0%), wheat flour (3.0%) and wholemeal wheat flour (4.0%), data set out in current legislation for flour products (BRASIL, 2005). However, these figures show the quality of the bran obtained and that it can be used to prepare a variety of foods.

Moisture in bran shows the percentage of free water that exists in a sample in its original state and is responsible for the formation of lumps (ICTA/UFGRS and SILVA, 2013) and, in addition, lower moisture levels ensure greater product durability.

As a result, coconut bran was evaluated with an average value of 1.58%, which is within the standards established by RDC No. 263 of 2005, which states that the moisture content of bran should not exceed 15%. Some studies have also demonstrated the quality of the bran obtained in this work, when compared to algaroba flour (5.8%), prepared by Silva et al (2007), for dehydrated cassava bran Rodrigues (2010) verified a content of (12.01%), and Santos (2014) obtained a percentage of (2.96%) of moisture in licurí flour, corroborating the previous statements.

According to Brazilian legislation (Ordinance 354/1996), wholemeal flour should contain a maximum of 2.5% ash, but this study found higher values (4.8%), which is justified by the high mineral content and leads us to further studies to determine the components present in the composition.

Table 7 below shows the data obtained for proteins, lipids, fibre and reducing sugars.

Table 7: Average results for the protein, lipid, fibre and reducing sugar parameters.

Parameters			
Sample	**Protein (%)**	**Lipid (%)**	**Fibre (%)**	**Total Soluble Sugars (%)**
Mesocarp bran	1,82 ±0,16	2,30 ±0,47	45,50 ±1,97	13,65 ±0,00

Averages of three repetitions followed by the respective standard deviations.

The percentage of protein in coconut bran was not very considerable (1.82%), but when compared to cassava bran (1.50%) the result was satisfactory, as it had a small percentage increase. Brazilian legislation (RDC 263/2005) adopts protein content values for wheat and cassava flour, which cannot contain less than 7% and 1.50%, respectively. Studies carried out by Dias et al (2006), Neto (2003) and Chisté (2003) on cassava flour showed protein contents of 1.50%, 2.60% and 1.38% respectively, with Chisté's (2003) data being outside the established limits.

As for lipid content, this study found a value of 2.3%. This level is quite considerable when compared to the studies by Neto (2003), Dias et al (2006) and Chisté (2010) on cassava flours, which obtained lipid values of 0.91%, 1.39% and 1.02% respectively. Silva (2013) found values closer to this work when he obtained lipid levels of 1.80% in the mesocarp of the marizeiro fruit *in natura* and 1.40% in the cooked marizeiro mesocarp.

The insoluble fibre content of coconut mesocarp bran was 45.5%. It can be considered that bran is a very rich source of hemicellulose, which helps to absorb water and increase the faecal bolus, reducing transit time through the intestine (CECCHI, 2003). Other studies have found values lower than those found in this study, such as Bernardino (2011), Carvalho et al (2006) and Freitas et al (2008), with respective values of 31.3%, 20.68% and 22.22%.

5.3   Physico-chemical characterisation of cupcakes enriched with coconut mesocarp and without mesocarp.

The cupcake formulations made with different concentrations of coconut mesocarp bran and the control are shown in Figure 11.

Cupcake formulations made with different concentrations of coconut mesocarp bran.

Figure 11: Cupcake formulations made with different concentrations of coconut mesocarp.

Source: Own authorship.

After preparation, the samples were subjected to physico-chemical analyses in order to assess the nutritional value of the cupcakes enriched with coconut mesocarp. The data found for pH, acidity, ash and moisture of the cupcake formulations with different proportions of mesocarp bran are shown in Table 8.

Table 8: Average results for pH, acidity, ash and moisture.

Samples	Parameters			
	PH	Acidity (%)	Ash (%)	Humidity (%)
Control	8,71 ± 0,24[a]	0,56 ± 0,19[a]	1,15 ± 0,04[a]	31,54 ± 1,35[a]
1%	8,29±0,18[ab]	0,38 ± 0,005[a]	1,02±0,04[b]	28,44 ± 3,07[a]
3%	8,20 ± 0,08[b]	0,37 ± 0,01[a]	1,03±0,05[b]	28,97 ± 2,29[a]

Averages from three replicates followed by the respective standard deviations. Equal letters in the same column do not differ according to Tukey's test at 5% significance.

The data above shows that there were no significant differences in the acidity of the samples (Control, 1% and 3%), but these values are directly related to the pH of the bran, which did not interfere with the acidity of the product; the higher the percentage of mesocarp added will not influence the acidity of the cake. Muniz (2014) evaluated the acidity of the addition of algaroba vargens flour and found that the percentage ranged from 6.27 to 6.70%, which is higher than in this study.

However, the pH values are close to those found by Bernardino (2011) who studied the pH of cake enriched with sugar cane bagasse flour, whose values ranged from 5.48 to 6.65. The addition of bran to the cupcake did not contribute to an increase in acidity, thus making the product ideal for consumption without causing any health problems.

According to Table 10, there was a significant difference between the samples in terms of ash content. This difference can be seen in the 1% and 3% samples, which differ from the Control, but do not differ from each other. A study by Maia 2007 found an ash content of 1.88% in cake formulations with passion fruit flour.

Another important item is the study of humidity, which is totally linked to food preservation, as the higher the percentage of humidity, the greater the perishability. The cupcake samples did not differ according to Tukey's test (p<0.05), ranging from 28.44 to 31.54 per cent. Reducing humidity ensures a longer shelf life for the product and prevents deterioration of the product by spoilage microorganisms. For Maia (2007) and Bernardino (2011) in their research into adding passion fruit flour and sugar cane bagasse flour to cakes, the variations in humidity were 33.08, 14.70 and 15.93% respectively. These results indicate that it is possible to produce source cakes with an increase in fibre with similar technological characteristics (MAIA, 2007). Table 9 below shows the protein, lipid and fibre values for the cupcake samples with different formulations.

Table 9: Average results for proteins, lipids and fibre.

Samples	Parameters		
	Protein (%)	Lipid (%)	Fibre (%)
Control	3,59 ± 0,53[a]	5,43 ± 0,57[a]	0,26±0,17[b]
1%	4,90 ± 0,33[a]	4,99 ± 0,07[a]	0,54 ± 0,06[b]
3%	4,11 ±079[a]	5,93 ± 0,13[a]	1,13 ± 0,07[a]

Averages from three replicates followed by the respective standard deviations. Equal letters in the same column do not differ by the Tukey test at 5% significance.

In terms of protein levels, the samples did not differ from each other and the values found varied very little between the samples, i.e. the more bran added to the product, the higher the protein percentage. This compares with the study by Maia (2007) who, when making cakes with the addition of passion fruit flour, found protein values in the range of 3.07 to 4.70 per cent. Muniz et al (2014) in their research into cake production with algaroba shell flour obtained values ranging from 7.63 to 7.84%, this variation was significant for the study and therefore higher than the study with coconut mesocarp.

There was no significant difference in lipid content between the cupcakes analysed, with averages of 5.43%, 4.99% and 5.93% respectively. Enriching the cupcake with mesocarp did not contribute to the increase in lipids in the cakes. Carneiro (2015) in studies on the characterisation of cakes with partial substitution of wheat flour with oats, quinoa and linseed obtained percentages ranging from 6.21 to 7.54%, however these values are higher in relation to the study of coconut mesocarp.

The percentage of fibre in the cupcake showed a significant increase in the samples: the greater the addition of mesocarp, the higher the percentage of fibre in the product, as shown by the fibre analysis and the statistical method. The variation was from 0.26 to 1.13% fibre in the cupcakes respectively.

5.4　Microbiological characterisation of cupcakes enriched with coconut mesocarp bran

Table 10 shows the results of the microbiological analyses of the cupcake formulations with different concentrations of green coconut mesocarp.

Table 10: Microbiological analysis of cupcakes enriched with coconut mesocarp and without it.

Parameters	Samples			
	Control	1%	3%	VMP
Coliform at 35 °C (MPN/g)	<3	<3	<3	-
Coliform at 45 °C (MPN/g)	<3	<3	<3	$5x10^2$
Coagulase positive staphylococci	Absent	Absent	Absent	-

(CFU/g)				
Moulds and yeasts (CFU/g)	Absent	Absent	Absent	-
Salmonella sp/25g	Absent	Absent	Absent	Absence
Escherichia coli	Absent	Absent	Absent	-

[1]VMP: Maximum permitted value, according to Brazil (2001).

[2]MPN: Most Probable Number.

[3] CFU: Colony forming unit.

Considering the results obtained after the microbiological analyses, it can be seen that the cupcakes produced with the addition of mesocarp and the control met current legislation in terms of the number of thermotolerant coliforms and *Salmonella* sp/25g. The absence of these microorganisms in the samples analysed indicates that the raw materials and their respective processes were efficient, as food contaminated by these bacteria is considered a source of human contamination, posing a great risk to public health.

Moulds and yeasts, *Escherichia coli* and Staphylococus *coagulase* positive, their absence in the food shows that good manufacturing practices were used efficiently during the processes, with no contamination by the handler or cross-contamination. The coliform values presented at 45 °C indicate that the cupcakes do not pose a risk to health, that there was no major contamination and that the food was handled under good hygiene conditions and good manufacturing practices were used. During an experiment using Licuri (Syagrus *coronata)* flour added to cakes and biscuits, Santos 2014 obtained similar values, which were below the levels tolerated by the legislation (BRASIL, 2001).

1.5   Sensory evaluation of the cupcakes made

1.5.1 Acceptance test

The average scores, with their respective standard deviations, given by the tasters to the cupcakes in relation to the attributes appearance, colour, aroma, texture, flavour and overall acceptance are shown in Table 11.

Table 11: Acceptance test results for cupcakes with coconut mesocarp bran added in different concentrations.

Attributes	Formulations		
	control	**1%**	**3%**
Appearance	8,06 ± 1,38[a]	8,04 ± 1,07[a]	7,89 ± 1,47[a]
Colour	8,02 ± 1,26[a]	7,89 ± 1.12[a]	7,77 ± 1,45[a]
Flavour	7,64 ± 1,46[a]	7,88 ± 1,11[a]	7,95 ± 1,17[a]
Flavour	7,63 ± 1,47[b]	7,65 ± 1,30[b]	8,11 ± 1,17[a]
Texture	7,17 ± 1,83[b]	7,36 ± 1,44[ab]	7,69 ± 1,52[a]
Global acceptance	7,60 ± 1,44[a]	7,58 ± 1,23[a]	7,94 ± 1,25[a]

Purchase intention	2,04 ± 1,14[a]	1,97 ± 1,06[a]	1,79 ± 1,00[a]

Control: (0% mesocarp meal); 1%: (1% mesocarp meal); 3%: (3% mesocarp meal). Averages from 146 repetitions followed by the respective standard deviations. Equal letters on the same line do not differ according to Tukey's test at 5% significance.

Table 11 shows that the averages of the experiments for the different attributes analysed did not differ statistically from each other at the ($p<0.05$) significance level. However, the average values of the attributes ranged from 7.17 to 8.11, equivalent to the hedonic terms "extremely disliked" to "extremely liked". Among the formulations, the one with 3% coconut mesocarp bran showed better acceptance of the attributes evaluated. The 3% formulation differed significantly from the control and 1% samples in terms of flavour. The texture attribute showed a significant difference between the 3% and control samples, but did not differ from the 1% sample.

According to Carderelli (2006), researchers developing new products need to know not only the degree of overall acceptability, but also what consumers like or dislike about the product, and how these attributes can be modified to increase acceptability. In this way, studies often include questions about product attributes that can determine the level of overall acceptability and questions related to the properties of the food, such as aroma, flavour and texture. The following are the scores given by tasters in relation to the appearance attribute of cupcakes in different formulations.

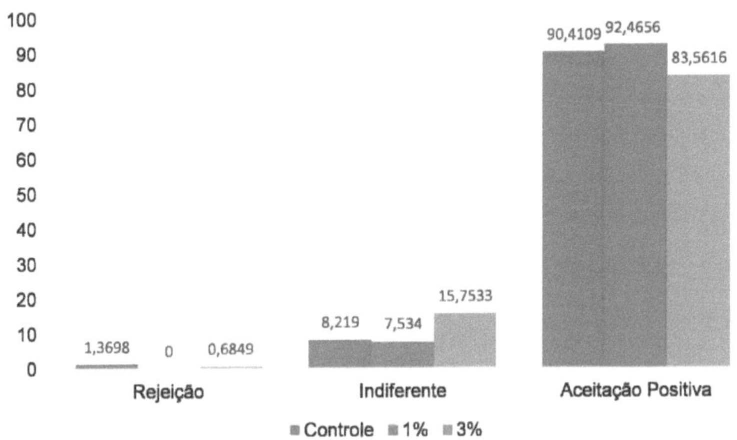

Figura 12: Frequency histograms for the appearance attribute of cupcakes enriched with coconut mesocarp bran.

When evaluating appearance, a wide range of characteristics are actually investigated, such as: colour (the most important factor in appearance): product dimensions, type of cut, etc. This parameter is one of the overall acceptance attributes of a product. The Control, 1% and 3% samples obtained excellent acceptance, as shown in the figure above, with no significant difference between the products. The addition of bran to the cupcakes did not interfere in this respect, i.e. the higher the percentage of bran added, the better the acceptance of the product, as it did not interfere in the results.Figure 13 shows the scores given by the judges in relation to the colour attribute of the cupcakes enriched with coconut mesocarp.

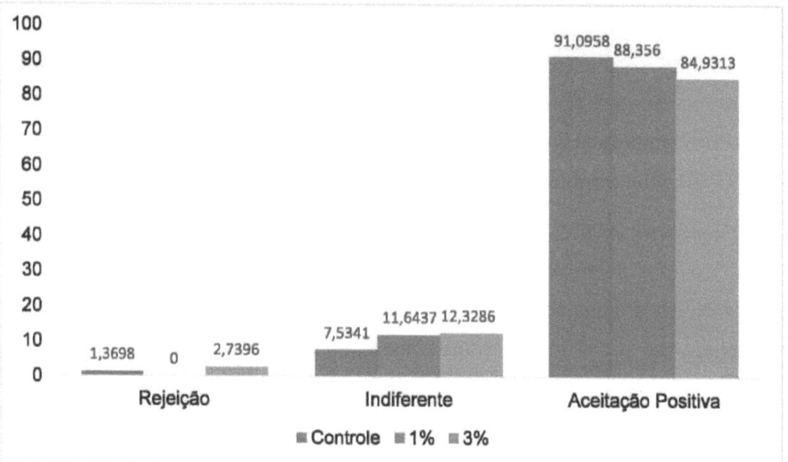

Figura 13: Frequency histograms for the colour attribute of cupcakes enriched with coconut mesocarp bran.

The samples showed excellent acceptance ratings for the Colour attribute with values of 91.0958%, 88.356% and 84.9313% respectively. The addition of mesocarp bran had a positive effect on this attribute. The rejection percentages were very low, characterising good acceptance by the evaluators. Figure 14 shows the scores given by the evaluators in relation to the aroma attribute of the cupcakes with different formulations.

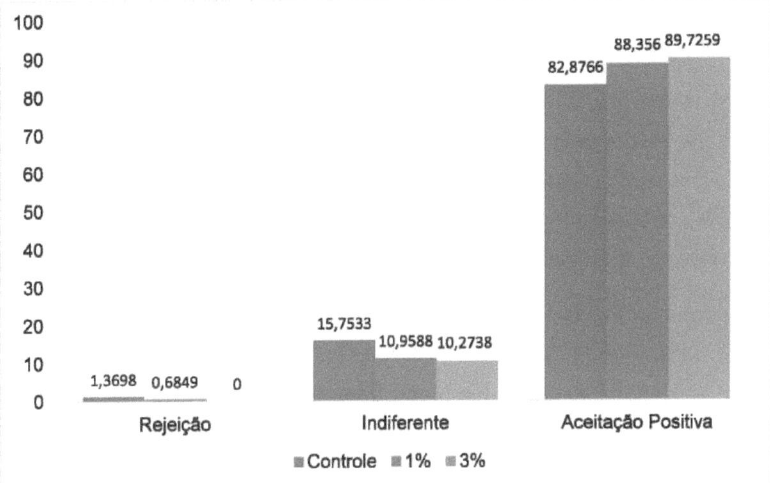

Figura 14: Frequency histograms for the flavour attribute of cupcakes enriched with coconut mesocarp bran.

With regard to flavour, what is assessed is the set of volatile aromatic compounds that are naturally released by the products. Spoilage processes, both enzymatic and caused by microorganisms, end up releasing characteristic aromatic compounds that are easily detected by smell (HERNANDES ET AL 2007).

The Control, 1% and 3% formulations, according to the acceptance test, showed excellent levels of acceptability, with little rejection by the evaluators during the test.

Figure 15 shows the judges' scores for the flavour attribute of cupcakes in different formulations.

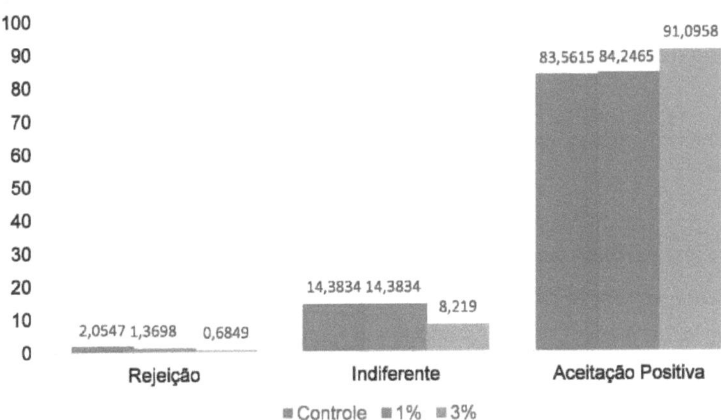

Figura 15: Frequency histograms for the flavour attribute of cupcakes enriched with coconut

mesocarp bran.

According to the results obtained for the flavour attribute, the Control samples (83.5615%), 1% (84.2465%) and 3% (91.0958%) showed no significant difference for this parameter and were well accepted by the evaluators. The rejection rate was very low according to the figure above (4%), with 32% of the evaluators reporting that the samples were indifferent in terms of flavour.

Flavour plays an important role in influencing the consumer's choice of a particular product (BURITI, CIRDARELLI, SAAD, 2008). It can therefore be seen that the higher the concentration of coconut mesocarp in the product, the better the flavour, which may be associated with the concentration of sugars present in the coconut mesocarp bran.

Figure 16 shows the scores given by the tasters in relation to the texture of the cupcakes at different concentrations.

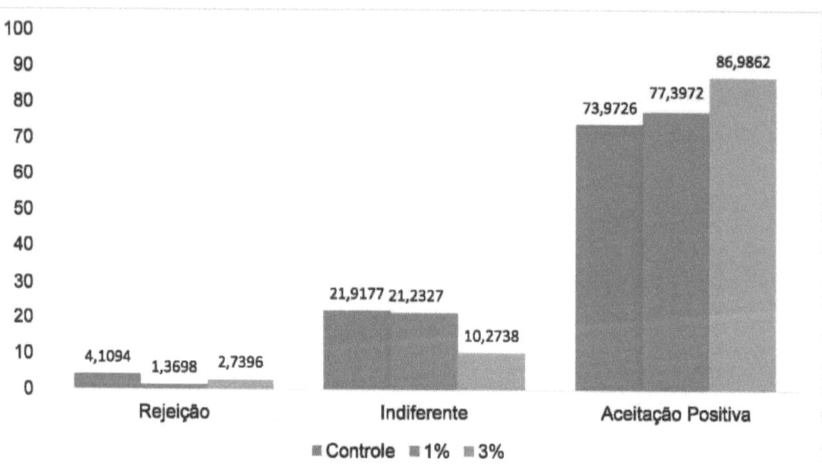

Figura 16: Frequency histograms for the texture attribute of cupcakes enriched with coconut mesocarp bran.

With regard to texture, there was no difference between the samples for this parameter. The cupcakes added with mesocarp showed excellent acceptance, but did not differ from the control sample in this respect. Figure 17 shows the scores given by the tasters in relation to overall acceptance.

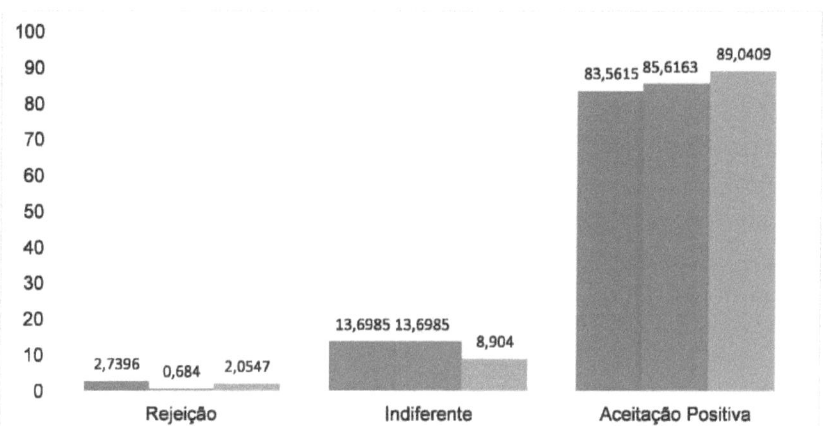

Figura 17: Frequency histograms for the overall acceptability attribute of cupcakes enriched with coconut mesocarp bran.

The cupcakes produced showed excellent acceptance, ranging from 83.55 to 89.03% between the formulations, characterising excellent acceptance of the products produced. The formulation with the highest acceptance rate was 3% with 89.03%, followed by 1% with 85.06% and 83.55% for Control. This shows that the addition of mesocarp bran did not alter the attributes studied in this research, improving their sensory aspects.

5.5. 2Purchase intention test

The following (Figure 18) shows the scores given by the tasters in relation to their intention to buy the cupcake samples made with different proportions of mesocarp bran.

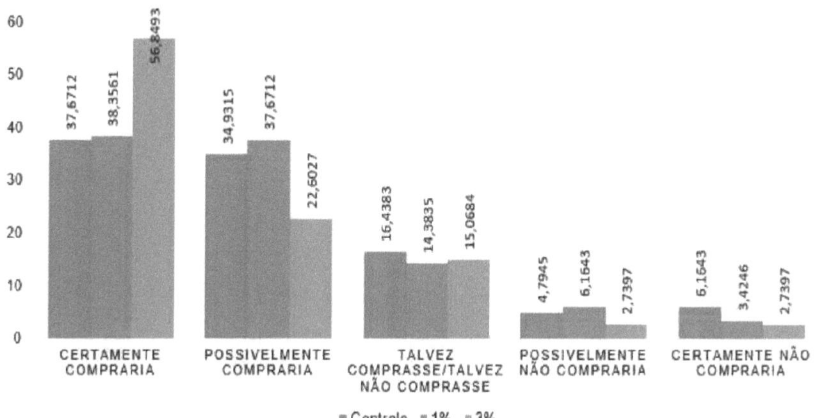

Figure 18: Percentage intention to buy cupcakes enriched with different proportions of mesocarp bran.

Regarding the intention to buy the cupcakes, 79.44% of the tasters were interested in purchasing the 3% coconut mesocarp bran formulation, which was the most acceptable, followed by the 1% sample with 76.02% of tasters intending to buy. The Control sample had a lower purchase intention than the others at 72.06%. The absence of bran in the Control sample resulted in lower acceptance by the tasters, while the addition of bran improved the sensory conditions of the products.

According to Cardarelli (2006), a product that doesn't score well in consumer acceptance tests will probably fail when it goes on sale, as the organoleptic characteristics usually come first for the consumer.

CHAPTER 6

Conclusion

The bran from the mesocarp of the coconut (Cocos *nucifera* L.) is an excellent option for adding to foods, considering the results obtained in this study, which showed a very high fibre content and a fairly considerable fat value.

The cupcakes made were within the microbiological standards according to current legislation.

Cupcakes enriched with coconut mesocarp bran are characterised as a functional food because they are a source of fibre, protein and minerals due to the ingredients in their formulations. Due to the high fibre content of the bran, the enriched samples are an interesting proposal for the high-fibre bakery products market.

The addition of coconut mesocarp made the product more attractive and with better sensory characteristics, but among the formulations evaluated by the judges, the one with the highest acceptance was the formulation with 3% coconut mesocarp bran.

CHAPTER 7

Bibliographical reference

(ABIMA) Associação Brasileira das Indústrias de Massas Alimentícias e Pãos&Bolos Industrializados, available at: <http://www.abima.com.br/noticias_eabima.php?id=765br/noticias_eabima.php?id=765 >. Accessed on: 09 June 2016.

Association of Official Analytical Chemistry - AOAC. Official methods of analysis. 15 ed. Washington, 1995.

ABIMAPI. Brazilian Association of Pasta Industries and Industrialised Breads & Cakes. Available at: <http://www.abimapi.com.br/estatistica-paes-bolos.php>. Accessed on: 15 August 2016.

ATHIÉ, I, PENTEADO, M., GOMEZ, VALENTINI, S. Grain conservation. Campinas, SP; Cargill Foundation, 236p, 1998.

BNDS. National Bank for Sustainable Development. Innovation in the food industry: dynamic importance in the Brazilian agro-industrial complex. Sector 37, p.333-370, 2015.

BASHO, S. M.; BIN, M. C. Properties of functional foods and their role in the prevention and control of hypertension and diabetes. INTERBIO, v.4, n.1, 2010. - ISSN 1981-3775.

BEZERRA, V. S. Dry and mixed cassava flours. Brazilian Agricultural Research Corporation - Brasília, DF: Embrapa Informação Tecnológica, 2006. 44 p.; (Family Agroindustry).

BERNARDINO, M. A. Characterisation and application of sugar cane bagasse flour in cakes. 2011. 85 f. Dissertation (Master's in Food Engineering) University of São Paulo, Pirassununga, 2011.

BRAZIL, Ministry of Health. National Health Surveillance Agency. Resolution no. 12 of 2nd January 2001. Available at: <http://www.anvisa.gov.br/legis/resol/12_01rdc.htm>.

BRAZIL. Ministry of Agriculture, Livestock and Supply. Normative Instruction No. 62 of 26 August 2003. Official Analytical Methods for Microbiological Analyses for the Control of Products of Animal Origin and Water. Official Gazette of the Federative Republic of Brazil, Brasília, 18 Sep 2003.

BRAZIL. Resolution RDC no. 263 of 22 September 2005. Approves the Technical Regulations for cereal products, starches, flours and meals set out in the Annex to this Order. Federal Official Gazette, Brasília, DF, 23 September 2005.

BURITI, F. C. A.; CARDERELLI, H. R.; SAAD, S. M. I. Instrumental texture and sensory evaluation of symbiotic fresh cream cheese: implication of the addition of Lactobacillus paracasei and inulin. Revista Brasileira de Ciências Farmacêuticas, vol. 44, n. 1, jan./mar., 2008.

CARDARELLI, H. R. Development of symbiotic petit suisse cheese. São Paulo, 133, 2006.

CARDOSO, T. G.; CARVALHO, V. M. Food poisoning by Salmonella spp. Revista Inst. Ciência e Saúde, 2006; 24(2): 95-101.

CARNEIRO et al, Physico-chemical characterisation of cakes with partial replacement of wheat flour with oats, quinoa and linseed. Enciclopédia Biosfera. Centro cinetifico conhecer - Goiania, v. 11 n. 21; p. 3348, 2015.

CARRIJO, O. A; LIZ, R. S; MAKISHIMA, N. Green coconut shell fibre as an agricultural substrate. Horticultura Brasileira, Brasília - DF, v. 20, n. 4, p. 533-535, 2002.

CECCHI, H. M. Fundamentos teóricos e práticos em análise de alimentos. 2.ed. Campinas: UNICAMP, 2003.

CHAVES, M. C. V.; GOUVEIA, J. P. G.; ALMEIDA, F. A. C. ; LEITE, J. C. A.; SILVA, F. L. H. Physico-chemical characterisation of acerola juice. Revista de Biologia e Ciência da Terra, v. 4, n. 2, 2004.

CHISTÉ, R. C.; COHEN, K. O. Physico-chemical characterisation of manioc flour from the d'água group sold in the city of Belém, Pará. Revista Brasileira de Tecnologia Agroindustrial, v. 04, n. 1, p. 91-99, 2010.

COELHO, M. A. Z.; LEITE, S. G. F.; ROSA, M. F.; FURTADO, A. A. L. Utilisation of Agroindustrial Waste: Production of Enzymes from Green Coconut Shells. B.CEPPA, Curitiba, v. 19, n. 1, p. 33-42, jan./jun. 2001.

CORRANDINI, E.; ROSAS, M. F.; MARCEDO, B. P.; PALADIN, P. D.; MATTOSO, L. H. C. Chemical composition, mechanical and thermal properties of fruit fibre from green coconut cultivars. Revista Brasileira de Fruticultura, Jaboticabal - SP, v. 31, n. 3, p. 837-846, sep 2009.

DAMORADAN, S. Food chemistry. Porto Alegre: Artmed, 2010. DAMASCENO, K.S. F. S. C.; ALVES, M. A.; FREIRE, I. M. G.; TORRES, G. F.; AMBRÓSIO, C L. B.; GUERRA, N. B. Hygienic-sanitary conditions of "self- Services" around UFPE and the raw salads they serve. Hig. Aliment. 2002; 16(102/103): 74-8.

DEODATO, J. N. V. Production and microbiological, physicochemical and toxicological evaluation of Pilosocereus chrysostele flour and its use in black bread. 2015. 110 f. Dissertation (Master's in Agroindustrial Systems) Federal University of Campina Grande,

Pombal, 2015.

DIAS, L. T.; LEONEL, M. Physico-chemical characterisation of cassava flours from different locations in Brazil. Ciênc. Agrotec., Lavras, v. 30, n.4, p. 692-700, 2006.

DREHER, M. Food Sources and Uses of Dietary Fiber. Complex Carbohydrates in Foods. Mareei Dekker, 1999.

EMBRAPA. Coconut Production and Commercialisation in Brazil in the Face of International Trade: Panorama 2014. Editor Carlos Roberto Martins and Luciano Alves de Jesus Júnior, 2013.

EMBRAPA. Coconut Production and Commercialisation in Brazil in the Face of International Trade: Panorama 2010. Editor Carlos Roberto Martins and Luciano Alves de Jesus Júnior, 2011.

EMBRAPA. Brazilian Agricultural Research Company. FONTES, H. R.; RIBEIRO, F. E.; FERNANDES, M. F. (Ed.). Coconut, production, technical aspects. Brasilia, DF: Embrapa Informação tecnológica, 2003. 106 p. (Frutas do Brasil, 27).

FAO 2014. World Production. Available at: <www.faostat.org.br>. Accessed on: 20 May 2016.

FARIAS, E. S.; GOUVEIA, J. P. G.; ALMEIDA, F. A. C.; BRUNO, L.A.; NASCIMENTO, J. Drying cajá in a fixed bed dryer. In: CONGRESSO BRASILEIRO DE FRUTICULTURA, 18., 2002, Belém. Proceedings... SBF: Belém, 2002. CD.

FERREIRA, P. R. B et al. Physico-chemical characterisation of babassu (Orbignya sp) mesocarp from regions of Piauí. Proceedings of the XIX UFPL Scientific Initiation Seminar 20 to 22 October 2010.

FOALE, M.; HARRIES, H. Farm and Forestry Production and Marketing Profile for Coconut (Cocos nucifera). In: ELEVITCH, C. R. (Ed.). Specialty Crops for Pacific Island Agroforestry, Holualoa, Hawai'i: Permanent Agriculture Resources (PAR), 2009. Available at: <http://agroforestry.net/scps>. Accessed on: 18 Dec. 2010.

FRANCISCO JR., W.E. Biochemistry in High School (De) Limitations from the analysis of some Chemistry textbooks. Ciência & Ensino, 2008b.

FRANCO, B. D. G. M. Intrinsic and extrinsic factors that control microbial growth in food. IN: Landgraf M, FRANCO BDGM. Food microbiology. São Paulo: Atheneu; 1996. p. 13-25.

BRAZILIAN FRUIT. Coconut. Post-harvest. Technical editor Wilson Menezes Aragão. Brasilia. Embrapa Information Technology, 2002, 76p.

GUTKOSKI, L.C., ANTUNES, E" ROMAN, LT. Evaluation of the degree of extraction of wheat and maize flours in a colonial-type mill. Boletim Ceppa, Curitiba, v.17, n.2, p. 153-166, 1999.

HERNANDES, N. K. et al Sensory acceptance tests of red beetroot (Beta vulgaris ssp. vulgaris L.) cv. Early Wonder, minimally processed and irradiated. Revista de Ciências Tecnologia Alimentos, Campinas, vol. 27, p. 64-68, Aug. 2007.

IBGE. Municipal Agricultural Production. Available at: <htttp:www.sidra.ibge.gov.br/bda/pesquisa>. Accessed on: 20 March 2016.

INSUMES. Dietary fibre: its role in health. Available at " http://www.insumos.com.br/funcionais_e_nutraceuticos/materias/171 .pdf" Accessed 15 May 2016.

JAY, J. M. Modern food microbiology. 4th ed. New York: Van Nostrand Reinhold; 1992.

KAJIYAMA, T.; PARK, K. J. Influence of initial feed moisture on drying time in an atomising dryer. Revista Brasileira de Produtos Agroindustriais, v.10, n.1, p.1-8, 2008.

KRAUSE, M. V.; MAHAN, L. K. Food, nutrition and diet therapy. 11ª Edition. São Paulo: Livraria Roca, p.981. 2005.

LACERDA, D. B. C.L.; JÚNIOR, M. S. S.; BASSINELLO, P. Z.; CASTRO, M. V. L.; SILVA-LOBO, V. L.; CAMPOS, M. R. H.; SIQUEIRA, B. S. Quality of raw, extruded and parboiled rice bran. Tropical Agricultural Research. Goiânia, v. 40, n 4, p. 521-530, Oct./Dec. 2010.

LACERDA, D. B. C.L. Stability and quality of rice bran under different treatments and application of the extruded product in biscuits. 2008, 100 f. Dissertation (Master's in Food Science and Technology) Universidade de Goiáis, Goiânia, 2008.

MAIA, S. M. P. C. Application of passion fruit flour in the processing of corn and oat cake for special purposes. 2007, 90 f. Dissertation (Master's in Food Technology) Federal University of Ceará, Fortaleza, 2007.

MATTOS, A.L.A.; ROSA, M. de F.; CRISÓSTOMO, L.A.; BEZERRA, F.C.; CORREIA, D.; VERAS, L.de G.C. Processing green coconut shells. Fortaleza: Embrapa Agroindústria Tropical, 2011. 37p. il. Handout Available at: <http://www.ceinfo.cnpat.embrapa.br/arquivos/artigo_3830.pdf>.

MARTINS, A.; CUNHA, M. L. R. S. Methicillin resistance in Staphylococcus aureus and coagulase-negative Staphylococci: epidemiological and molecular aspects. Microbiol Immunol, v. 51, n. 9, p. 787-95, 2007.

MARTINS, L.T. Staphylococcus. In: TRABULSI, L.R. et al. Microbiologia. 3. ed. São Paulo: Atheneu, 1999. Gap. 18; p. 149-55.

MATSAKIDOU, A., BLEKAS, G" PARASKEVOPOULOU, A. Aroma and physical characteristics of cakes prepared by replacing margarine with extra Virgin olive oil. LWT - Food and Science Technology, Virginia, v. 43, n. 6, p. 948 - 957, Feb. 2010.

MENEZES, Adriana Carla Santos. Development of a fermented milk drink based on whey and cajá pulp (Spondia *mombim* L.) with potential probiotic activity. Thesis (Master's degree in food science and technology) - Federal Rural University of Pernambuco, Recife, 2011. Available at: http://www.pgcta.ufrpe.br/files/dissertacoes/20111 /Adriana_Carla_Santos_Menezes.p df Accessed: November 2011.

MIGLIATO KF, Moreira RRD, Mello JCP, Sacramento, LVS, Corrêa MA, Salgado HRN Quality control of the fruit of *Syzygium cumini* (L.) Skells. Rev Bras Farmacogn v. 17, p. 94-101.2007.

MORAIS, E. F.; MANIGLIA, E. B.; OMAE, J. M.; SOARES, L. F. F.; MADRONA, G. S. Development and evaluation of a cake based on carob *(Ceratonia siliqua)* flour. GEINTEC Magazine. São Cristóvão/SE - 2014. Vol. 4/n.5/ p.1340 - 1350 1340.

MOSCATTO, J.A.; PRUDÊNCIO-FERREIRA, S. H.; HAULY, M.C.O. Yacon flour and inulin as ingredients in chocolate cake formulation. Revista Ciência e Tecnologia de Alimentos, Campinas, v. 24, n. 4, p. 634-640, Oct.-Dec. 2004.

NATIONAL MUSEUM. Botanical Garden. National Museum of the Federal University of Rio de Janeiro. Accessed on 10 December 2015. Available at <http://www.museunacional.ufrj.br/hortobotanico/paginas/palmeiras/cocosnucifera.htm>

MUNIZ et al. Elaboration, physico-chemical and sensory characterisation of cake formulated with algaroba vargens flour. XX Brazilian Congress of Chemical Engineering - COBEC. Available at: http://pdf.blucher.com.http://pdf.blucher.com.br.s3-sa-east-1.amazonaws.com/chemicalengineeringproceedings/cobeq2014/1209-20471- 148223.pdf. Accessed on: 21 Sep 2016.

NETO, C. J. F.; FIGUEIREDO, R. M. F.; QUEIROZ, A. J. M. Physico-chemical evaluation of cassava flours during storage. Brazilian Journal of Products Agroindustriais, Campina Grande, v. 5, n. 1, p. 25-31, 2003.

NING, L.; VILLOTA, R.; ARTZ, W. E. Modification of Fiber Throwgh Chemical Treatments in Conernation. Cereal Chamisty. 1991.

OLIVEIRA, H. P. S., O consumo de alimentos funcionais - atitudes e comportamentos. Dissertation (Master's Degree in Communication Sciences, specialising in Markintg and Strategic Communication) - Fernando de Pessoa University, Porto, 2008.

OLIVEIRA, M. N.; SILVIERI, K.; ALEGRO, J. H. A.; SAAD, S. M. I. Technological aspects of functional foods containing probiotics. Revista Brasileira de Ciências Farmacêuticas, vol 38. n. 1, jan./mar., 2002.

ORDÓNEZ, J. A. Food technology: Food components and processes. Porto Alegre: Artmed, 2005.v1.

PARK, K. J.; ANTONIO, G. C.; OLIVEIRA, R. A.; PARK, K. J. B. Selection of Drying Processes and Equipment. Lecture, 1st August 2006. In: BRAZILIAN CONGRESS OF AGRICULTURAL ENGINEERING, 35th, 2006, João Pessoa. Proceedings..., João Pessoa: CONBEA, 2006. CD.

PARK, K. J.; ANTONIO, G. C.; Analyses of biological materials. Available at: <http://www.feagri.unicamp.br/ctea/manuais/analise_matbiologico.pdf>. Accessed on 15/10/2012.

PARSONS, C.M. Digestible amino acids for poultry and swine. Animal Feed Science and Technology, Netherlands, v.59, n.1, p. 147-153, 1996.

PEREIRA, C. L. Utilising green coconut waste to produce composites for rural construction. PhD thesis, Faculty of Animal Science and Food Engineering - USP. Pirassununga, 2012.

QUAGLIA, G. Ciência y Tecnologia de la Panificación. Zaragoza: Acribia, 1991.

PONTES JÚNIOR, V. A. Genetic Potential and Stability of Common Bean Families Obtained by Different Breeding Methods.
Dissertation (Master's Degree in Genetics and Plant Breeding) - Goiânia - GO: UFGO, 2012.

RAMOS; N. C,; PIEMOLINI-BARRETO, L. T.; SANDRI, I. G. Elaboration of pre-mix for gluten-free cake. Alimentos e Nutrição, Araraquara v. 23, n. 1, p. 3338, jan./mar. 2012.

RIBEIRO, E. P. SERAVALLI, E.A. G. Química de Alimentos, 2ª Ed. ver. São Paulo. Ed. Geral Blucher, 2007.

RODRIGUES, J. P. M. Characterisation and sensory analysis of starch biscuits enriched with cassava bran. 2010. 81 f. Dissertation (Master's in Food Science and Technology) Universidade Federal de Goiás, Goiânia, 2010.

ROSA, M. de F.; MATTOS, A. L. A.; CRISOSTOMO, L. A.; FIGUEIRÊDO, M. C. B. de; BEZERRA, F. C.; VERAS, L. de G. C.; CORREIA, D. Utilisation of green coconut shells. In: CARVALHO, J. M. M. (Org.). BNB support for research and development in regional fruit growing. Fortaleza: Banco do Nordeste do Brasil, 2009. chap. 8, p.165-190 (BNB. Ciência e Tecnologia, 4).

ROSAS, M.F. Alternativas para uso da casca do coco verde. Rio de Janeiro: EMBRAPA, 1998.

ROSTAGNO, H.S.; BUNZEN, S.; SAKOMURA, N.K.; ALBINO, L.F.T. Methodological advances in the evaluation of feeds and nutritional requirements for poultry and pigs. Revista Brasileira de Zootecnia, Viçosa, v.36, special supplement, p.295- 304, 2007.

SANTANA, I. A. Chemical and Functional Evaluation of Green Coconut Pulp Applied to Edible Ice Cream. Master's dissertation, Mauá School of Engineering, Mauá Institute of

Technology University Centre, São Caetano do Sul, SP, 2012.

SANTOS, M. H. O. Technological utilisation of lycurí (Syagrus coronata) processing waste, 2014. 61 f. Dissertation (Master's in Food Engineering) Universidade Estadual do Sudoeste da Bahia, Itapetinga, 2014.

SANTOS, M. H. O. Technological utilisation of waste from the processing of Licuri (Syagrus *coronata*). Dissertation presented to the Postgraduate Programme in Food Engineering at the State University of Southwest Bahia, Itapetinga - BA. UESB, 2014.

SANTOS, B. M. O.; AGUILLAR, O. M.; TAKAKURA, M. S. Simultaneous colonisation of Staphylococcus aureus in the nasal cavity and hands of healthy patients in a teaching hospital. Rev Microbiol, v. 21, n. 4, p. 309-14, 1990.

SANTOS, B. M. O.; SCOCHI, C. G. S.; SOUZA, M. T. G. Prevalence of healthy carriers of Staphylococcus aureus in nursing staff in paediatric units of a general teaching hospital. Part I. Rev Pauli Hosp, v. 38, p. 24-9, 1990.

SILVA, N.; JUNQUEIRA, V. C. A.; SILVEIRA, N. F. A.; TANIWAKI, M. H.; dos. SANTOS, R. F. S.; GOMES, R. A. R. Manual de métodos de análise microbiológica de alimentos e água, São Paulo, 4. ed. Livraria Varela, 2010.

SILVA, R. G. V. Physico-chemical characterisation of sweet potato flour for bakery products. Vitoria da Conquista - Bahia: UESB, 2010. 77p. (Dissertation - Master's in Food Engineering - Process Engineering), 2010.

SILVA, N. C. Sensory evaluation of a cookie containing babassu mesocarp flour. Monograph (Graduation in Food Engineering) - Bachelor's Degree in Food Engineering, Bom Jesus Advanced Campus/Federal University of Maranhão (UFMA), 2014.

SILVA, R. G. V. Physico-chemical characterisation of sweet potato flour for bakery products. Dissertation (Master's in Food Engineering) - Itapetinga-Ba: UESB, 2010.

SILVA, M. C. Evaluation of the microbiological quality of food using conventional methodologies and the SimPlante system. Master's dissertation from the Luiz de Queiroz College of Agriculture - ESALQ/USP. Piracibaca - SP, 2002.

SOUZA, P. D. J.; NOVENDO, D.; ALMEIDA, J. M.; QUINTILIANO, D. A. Sensory and nutritional analysis of savoury tart made from alternative uses of vegetable stalks and peels. Alimentos e Nutrição, Araraquara v. 18, n.1, p. 55-60, 2007.

SROAN, B. S.; BEAN, S. R.; MACRITCHIE, F. Mechanism of gas cell stabilisation in bread making. I. The primary gluten-starch matrix. Journal of Cereal Science, v. 49, p. 32-40, 2009.

STAINKI, R.D. The science of microbiology. 2012. Available at:<http://coral.ufsm.br/microgeralhttp://coral.ufsm.br/microgeral/Conteudo%20teorico/A%

20ciencia%20da%20micro biology.pdf>. Accessed 12/10/2012.

STONE.H.; SIDEL, J. L. Sensory evaluation practices. 3ª ed. London: Academic, 408p, 2004.

UFRGS. Federal University of Rio Grande do Sul. Evaluation of the technological quality of wheat flour. Available at: http://www.ufrgs.br/napead/repositorio/objetos/avaliacao-farinha-trigo/. Accessed on 24 April 2016.

VICENZI, R. Introduction to Food Analysis. Industrial Food Chemistry. Unijui, 2011.

YEMN, E. W.; WILLIS, A. J. The estimation of carbohydrate in plant extracts by anthrone. The Biochemical Journal, London, v. 57, p. 508-514, 1954.

ZANINI, C. D. et al. Physico-chemical evaluation of apple cake added with inulin among children. Disponovelem   :
<http://revistas.unincor.br/index.php/revistaunincor/article/downloadSuppFile/1105/92>. Accessed on: 15 Aug 2016.

CHAPTER 8

Appendix

Appendix A

8.1.1 Sensory evaluation form.

Name: Sex: F ( ) M ( ) Age:Date: //2016

You are receiving two coded samples of cupcakes enriched with coconut mesocarp bran in different proportions and a control sample. Please taste the samples, evaluating each of them on the attributes of: APPEARANCE, COLOUR, AROMA, TASTE, TEXTURE AND OVERALL ACCEPTABILITY. Mark the code for each sample in the table according to how much you liked or disliked the product.

(9) I liked it a lot
(8) I liked it moderately
(7) I liked it regularly
(6) slightly liked
(5) neither liked nor disliked

(1) extremely disliked
(2) moderately disliked
(3) regularly disliked
(4) slightly disliked

Attributes	Sample:	Sample:	Sample:
Appearance			
Colour			
Flavour			
Flavour			
Texture			
Global Acceptance			

Please now indicate how sure you are that you would or would not buy the cupcake samples you tried earlier.

1. I would certainly buy it
2. Would possibly buy
3. Maybe I would, maybe I wouldn't
4. Possibly wouldn't buy
5. I certainly wouldn't buy it

Sample no.	Value

3- Do you eat cupcakes? ____

4. How often do you consume? ____

8.2   Appendix B

8.2.1 Informed Consent Form

INFORMED CONSENT FORM

Research Title: Sensory evaluation of traditional cupcakes enriched with coconut mesocarp bran (Coco nucifera L.)

Responsible for the research: Luís Paulo Firmino Romão da Silva, Alfredina dos Santos Araújo and Everton Vieira da

Silva

You are being invited to participate as a volunteer in the sensory evaluation of traditional and enriched cupcakes.

The coconut (Coco *nucifera L.*) is botanically classified as a fibrous drupe, made up of: the epicarp or smooth epidermis, the layer that surrounds the mesocarp; the mesocarp, the thick, fibrous layer (shell); and the endocarp, the woody layer that surrounds the seed, making it very hard after ripening. Between the endocarp and the solid albumen is a thin layer, light in colour in immature fruit and brown in ripe fruit, called the integument. The albumen or solid endosperm in ripe fruit is a fleshy, white, oily, fairly thick layer, and in immature fruit, depending on its stage of ripeness, it has a semi-solid (gelatinous) consistency (EMBRAPA, 2003). The mesocarp fibre has the function of reinforcing materials, thanks to its rigidity and high resistance, giving products durability (ROSAS, 2009). As well as being a source of fibre, the mesocarp is also rich in minerals, proteins and has properties that can be used in the food industry as an important food supplement in the human or animal diet. It is extracted by crushing the mesocarp, drying it and then crushing and grinding it. After this process, it is introduced into food as a food supplement due to its richness in nutrients. The aim of this research is to carry out a sensory analysis of cupcakes enriched with coconut mesocarp (Coco *nucifera L.)*, using an acceptance test to assess parameters such as colour, taste, appearance, texture, aroma and overall acceptance of the product.

It is very unlikely that there will be any discomfort or risk for you who are taking part in the research. At the very least, an allergy to coconut mesocarp bran could occur. Both the cupcake and the mesocarp bran used have undergone microbiological analysis and did not pose any health risk. You will be informed about the research in any way you wish. You are free to refuse to take part, withdraw your consent or stop taking part at any time. Your participation is voluntary and refusing to take part will not result in any penalty or embarrassment. The researchers will treat your identity with professional standards of confidentiality. Participation in the study will be free of charge and no financial compensation will be available.

I, ID No., declare that I have read the information contained in this document, have been duly informed by the researcher of the procedures that will be used, risks and discomforts, benefits, cost/reimbursement of participants and confidentiality of the research. I also agree to take part in the research. I have been assured that I can withdraw my consent at any time without any penalty. I also declare that I have received a copy of this document and that I have had the opportunity to read it and clarify my doubts.

    Participant's signature   Signature of the Researcher in Charge

Name:
Date:
Contact:
Researchers:
Luís Paulo Firmino R. da Silva
Alfredina dos Santos Araújo
Everton Viera da Silva

Printed by Books on Demand GmbH, Norderstedt / Germany